SKY GATE

SKY GATE
ETIENNE DE MALGLAIVE

Airlife
England

Previous spread: Perfect symmetry between the wing of a Boeing 747-400 and an American Airlines' DC-10.

Copyright © 1999 Etienne de Malglaive

First published in the UK in 1999
by Airlife Publishing Ltd

British Library Cataloguing-in-Publication Data
A catalogue record for this book
is available from the British Library

ISBN 1 84037 019 X

The information in this book is true and complete to the best of our knowledge. All recommendations are made without any guarantee on the part of the Publisher, who also disclaims any liability incurred in connection with the use of this data or specific details.

All rights reserved. No part of this book may be reproduced or transmitted in any form or by any means, electronic or mechanical including photocopying, recording or by any information storage and retrieval system, without permission from the Publisher in writing.

Printed in Singapore

Airlife Publishing Ltd
101 Longden Road, Shrewsbury, SY3 9EB, England
E-mail: airlife@airlifebooks.com
Website: www.airlifebooks.com

◀ A Boeing Mirage? A 747 and a 737 taxi into Los Angeles in an evening heat haze.

▲ Sunrise at Paris CDG.

▼ These green recessed lights keep pilots on the straight and narrow when taxying between runway and terminal.

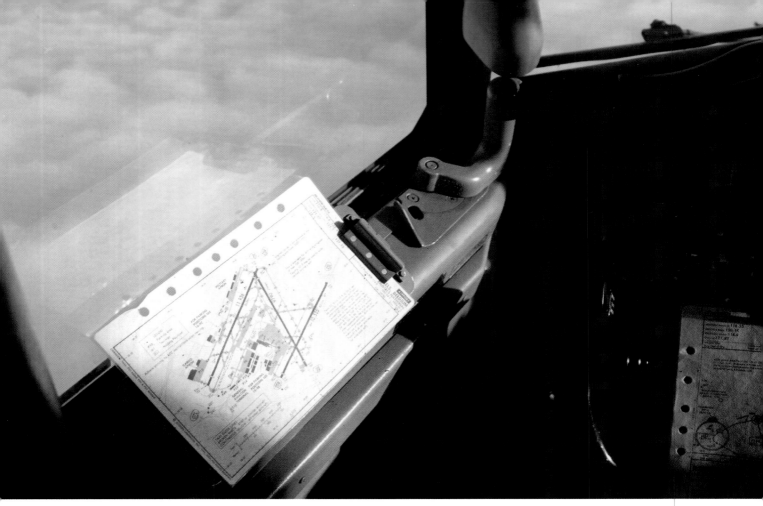

◀ Despite the electronic sophistication of the modern airliner, there is always the necessity of an up-to-date chart. This Jeppesen chart is conveniently clipped into the cockpit alongside the captain of a Boeing 737 as she descends to land at Brussels Airport.

▼ Turbo fan!

◀ This Saudia Lockheed L1011 TriStar 200 is about to loosen its last physical link with the terminal building before push-back and a lengthy flight.

◀A Boeing 747 under the pressure-jets of a de-icing station.

Following spread: **An Air France Boeing 767-300 ER on a dawn push-back at Paris CDG. This long-range twin jet will take longer to reach her destination than her fleet-sister Concorde, seen in the background – but will consume considerably less fuel.**

▼De-icing equipment rushes into action after a cold drizzle leaves aircraft on the apron at Montreal – Dorval covered in ice.

◀ An interesting study of the port main undercarriage of this Airbus A340. When the aircraft is empty it helps support 278,843 lb (126,481 kilograms) and when loaded 566,588 lb (257,000 kilograms).

▲ UTA was fully merged with Air France in 1990. Seen here at Paris CDG, this Boeing 747 is towed to her gate before loading.

▼ McDonnell Douglas DC-10s belonging to the French airline AOM. This new independent airline serves French islands in the Pacific and Caribbean.

Following spread: A Continental Airlines Boeing 737 descends on her approach to New York's Newark Airport.

◀ A Virgin Atlantic Airways Boeing 747 plugs into fuel lines before another ocean crossing. These aircraft can carry over 58,000 US gallons (229,000 lt).

Following spread: An Air France Concorde loads for a misty early-morning departure.

▼ A study in oils – topping up a FedEx MD-11.

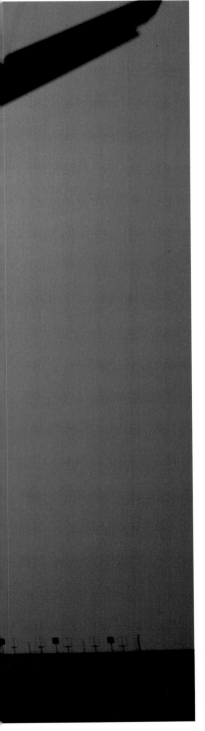

▼A foggy night in London town. An Airbus A340 at Heathrow.

Heavy traffic on the taxi-way at New York's JFK in the midday heat haze at this major international airport.

New York – JFK. A TWA Boeing 747 taxies out into the sunset for an overnight flight across the Atlantic.

▲ An Air France Boeing 747 undergoes one of its regular maintenance checks, shrouded in a web of scaffolding within an enormous hangar at Paris CDG.

◀ Airport traffic stops to allow an incoming Air France Boeing 747 to cross the road at Paris Charles de Gaulle.

▼ 150 million dollars of Boeing 747-400 is not going anywhere courtesy of a few cents' worth of timber.

Following spread: **Arrivals and departures at Paris Charles de Gaulle.**

◀ Air France mechanics close the stained hatch of a Boeing 747's auxiliary power unit.

▼ On days like this passengers appreciate modern boarding-gate facilities.

▼ The Washington Dulles terminal and tower reflect the 1960s mixture of grandeur and audacious simplicity.

▶ Coming in from the cold. Alaska Airlines' Boeing 737 sinks gently to earth against the sultry background of San Diego Bay.

◄ A Korean Airlines Boeing 747 on final approach into Los Angeles. The spectacular restaurant was built in 1961.

▲ A Boeing 747 tail fin seen through the exhaust haze of a US Air 737.

Following spread: Cargo Boeing 747s at Miami International Airport, unusually parked in a row like cars along the sidewalk.

◀ Only seconds from landing, this Boeing 747-400 of Thai Airways International arrives from Bangkok into Los Angeles LAX.

▲ Maintenance in progress on a Boeing 727.

◀ Tricks of the light as the extended flaps of a Boeing 737 cause atmospheric distortion before a setting sun.

▼ An Aer Lingus 747 takes aboard sufficient victuals to keep 480 passengers happy for eight hours.

▶ A touch of nostalgia. Pan Am as it was in 1988.

▼ A Laker DC-10 – the man who took on the giants of airline travel is back again.

◀ Checking in can take hours – followed by moments of sheer panic. ▼

Following spread: Flight delayed – the length of this Air France Airbus A340 is 208 feet and 9 inches (63 metres), giving a fair indication of the density of the fog on this early morning at Paris CDG.

◀ Scenic Cessna 172 at the tiny Goulding Airport in Utah. The spectacular Monument Valley is only two minutes' flight time away.

▼ Point of contact.

Following spread: A mighty 747-400 gets ready to roll as the last ground crew runs clear of the pulsing turbofans.

111

◀ FedEx loads this MD-11 at Paris CDG – and delivers to a remote gas station at Sunsites Arizona. ▼

▲Two Rolls-Royce turbofans wait to deliver 40,000 pounds of thrust to get this British Airways Boeing 757-200 airborne on her evening departure from Paris Charles de Gaulle.

▼Seen through the cobweb of Los Angeles Lax approach lights, a Jetstream 31 alights in the early morning light.

▲ 'Annule' – it's the same in any language!

◀ Not as alarming as it looks! French airline workers use distress flares in a 1995 strike and the riot police move in against a background of red smoke.

Following spread: **Symmetry of aerofoils.**

◀ The hard edges of Phoenix surround the smooth aerodynamic lines of an Alaska Airlines MD-83.

Following spread: Boeing introduced the 737 in 1967 as competition to the British BAC 1-11 and fellow American DC-9. It utilised the nose and fuselage sections of her famous 707/727 predecessors. It became the world's best selling airliner. The 767, seen here in close up, was designed to compete with the Airbus A310 and was conceived in 1978. Both are still in production.

◀ At home in Toronto, these Dash 8s were built by de Havilland Canada and are operated by Air Ontario, an associate carrier of Air Canada.

Following spread: **Pan Am Airbus A310 –** *Clipper Munich* **– returns to New York JFK at dusk.**

Following spread: A sudden deluge of rain from a tropical storm sends an airport driver scurrying for the shelter of a hangar at Kuala Lumpur's Sepang Airport.

Previous spread: Sunrise over Paris – Orly.

Hong Kong Airport – Chek Lap Kok.

▲A new generation of aviators inspect the state of the art at London – Gatwick.